Omelio Cepero Rodríguez
Loisy Valero Hernández
Zuleiny Meneses Martin

Environmental education from community approaches

Omelio Cepero Rodríguez
Loisy Valero Hernández
Zuleiny Meneses Martin

Environmental education from community approaches

Environmental education from community approaches

ScienciaScripts

Imprint
Any brand names and product names mentioned in this book are subject to trademark, brand or patent protection and are trademarks or registered trademarks of their respective holders. The use of brand names, product names, common names, trade names, product descriptions etc. even without a particular marking in this work is in no way to be construed to mean that such names may be regarded as unrestricted in respect of trademark and brand protection legislation and could thus be used by anyone.

Cover image: www.ingimage.com

This book is a translation from the original published under ISBN 978-620-3-58737-1.

Publisher:
Sciencia Scripts
is a trademark of
Dodo Books Indian Ocean Ltd., member of the OmniScriptum S.R.L Publishing group
str. A.Russo 15, of. 61, Chisinau-2068, Republic of Moldova Europe
Printed at: see last page
ISBN: 978-620-3-68493-3

TITLE: Environmental education from community approaches

Authors

Omelio Cepero Rodríguez Loisy Valero Hernández Zuleiny Meneses Martin

SUMMARY:

One of the fundamental purposes of environmental education is to ensure that individuals and communities understand the complex nature of the environment and acquire the knowledge, values and practical skills to participate responsibly and effectively in the prevention and solution of environmental problems and in the management of environmental quality.

Environmental education is key to understanding the existing relationships between natural and social systems, as well as to achieving a clearer perception of the importance of socio-cultural factors in the genesis of environmental problems. In this line, it should promote the acquisition of awareness, values and behaviors that favor the effective participation of the population in the decision-making process. Environmental education thus understood can and should be a strategic factor that influences the established development model in order to reorient it towards sustainability and equity. - To make local communities responsible and committed to the processes of change and social transformation, confronting their problems, needs and demands with the possibilities and limitations (geographic, demographic, infrastructural, economic, technological, etc.) of the reality of which they are a part, expanding their capacities for initiative and criticism without - as a matter of principle - renouncing the advantages that scientific knowledge and technological innovation can offer for the promotion of an increasingly autonomous and sustainable development.

Key words: communities; environmental education, community development.

INTRODUCTION

Environmental education and the community dimension

For Environmental Education, as for any other educational praxis that intends to assert itself as a social and cultural project, community development is a key reference (Caride, 1998): in it, education and environment are unequivocally integrated, trying to transfer self-confidence, protagonism and self-management capacity to local communities and the different social groups that articulate them, in order to turn them into subjects of the development process and not mere objects of it. In this sense, it is intended to recognize the aggregates of people not only as a sum of individuals but as a "community" inscribed in a territory, with a "common" past and future, from their daily life to their progressive integration in other communities and realities (regional, national, international), without renouncing to the best and most worthy conditions of their quality of life. As Unamuno expressed, before even glimpsing the stage of globalization in which we now live, "the human being becomes planetary when he recovers his self-control, his roots in his immediate territory, the identification of his nation and his local culture and then, finally, his role in the world. On the other hand, the convergence between Environmental Education and Community Development cannot ignore the fact that "communalism" has been one of the important ideological sources both for the articulation of environmentalist thought and for the construction of proposals that support local development (Boockhin, 1978; Schumacher, 1978; Bosquet, 1979). This is highlighted by Dobson (1997: 171) when he states "that a common problem with the strategy of lifestyle change is that it is ultimately detached from the end it seeks to achieve, in that it is not obvious how the individualism on which it is based will develop into the communitarianism that is central to most descriptions of the sustainable society". In fact, the author continues, "it would seem more sensible to subscribe to forms of political action that are already communitarian and that, therefore, are practical and at the same time a foretaste of the announced goal. In this way, the future is inserted into the present and the program becomes more intellectually convincing and practically coherent". Castells (1998: 141), makes a pronouncement in accordance with this perspective; for him, "the mobilization of local communities in defense of their space, against the intrusion of undesirable uses, constitutes the most rapidly developing form of ecological action and the one that perhaps links in a more direct way the immediate concerns of the people with the broader issues of environmental deterioration". To put it another way: local responses to environmental problems are testing forms of community reaction to global challenges.

Environmental Education will have to feed on this conviction and its congruence if it intends to play a significant role in the construction of community projects, although without falling into the illusionism of an idealized "community", thought of as a mythical and utopian space, as Furter (1983) warns us, when he notes that the values of community cohesion - to a large extent, latent in rural societies - have been subjected to profound and contradictory changes: the expansion of new productive models, the mixtification of the referents of collective identity due to the homogenizing impact of "global culture", the increase in the individualized perception of existence, the tendency to solve social problems in the sphere of privacy, the increase in forced mobility, the expansion of urbanism, the penetration of new communication and information technologies, etc.

Moreover, we must beware of the belief that micro-spaces simplify problems by reducing their dimension. If "small is beautiful" as Schumacher suggested, this does not mean that its level is less complex, more homogeneous or more consensual than others (Furter, 1983: 106). Using a metaphor suggested by Edgar Morin, in contemporary communities all the problems, contradictions and complexities of the globalized world are expressed as in the fragment of a hologram; even in those who are not aware of belonging to it. From this point of view, as expressed on the global scale of climate change - to give an example - the environmental crisis is part of the globalization process. Beyond these arguments, it is also worth noting the importance given in recent decades to the desirable integration of educational processes in the dynamics of each social reality, particularly those built *from*, *with* and *for* local communities (towns, neighborhoods, cities).

The Local Agenda 21 approach, based on the recommendations of the 1992 Rio Summit, is part of this trend. In general, it seeks to contribute to the creation of conditions of citizenship and social well-being that are increasingly congruent with the principles that inspire the harmonious, integral and sustainable development of each individual and each community. For Calvo and Franquesa (1998), after admitting that for years in the documents of different international events on Environmental Education an express allusion to democracy as a general framework or to the deepening of democracy as a concrete action was avoided for political reasons, today we can clearly state the option for equity as a principle and democracy as a suitable framework. Dialogue, participation, negotiation and consensus are thus the mechanisms for resolving conflicts or, at least, for raising them; and the involvement of people in these procedures is an essential part of their training.

In this sense, we can see how ideological, socio-political and methodological approaches that insist on claiming social development based on what is "common" to people, considering aspects as diverse as landscape, culture, feelings or experiences that are configured in a given territory, try to validate models and processes of community development that emphasize the possibilities of education in the achievement of three main objectives:

- To advance in the potentialities offered by promoting the reencounter of local communities with themselves, guaranteeing the survival of the territory and of the social groups (from children to the elderly) that inhabit it, including an adequate availability of its natural resources and respect for the values that take as reference the different manifestations of the artistic-cultural heritage bequeathed by previous generations. This requires the compatibility of local and global dimensions, the micro vision with the macro vision, the civil society with the State, self-esteem with the appreciation of the other, identity with the universal, etc. Because, as Bassand (1992: 116), for local development "the singular is not incompatible with the local, local initiative implies endogenous but also and above all exogenous solidarities; local projects are not opposed to openness and exchange with the world, the local does not exclude the global; the tradition in which the identity is often rooted does not reject modernity.

El escenario comunitario

¿Qué particularidades posee?

To make local communities responsible and committed to the processes of change and social transformation, confronting their problems, needs and demands with the possibilities and limitations (geographical, demographic, infrastructural, economic, technological, etc.) of the reality of which they form part, expanding their capacities for initiative and criticism without - as a matter of principle - renouncing the advantages that scientific knowledge and technological innovation can offer in the face of the promotion of an increasingly autonomous and sustainable development. Marchioni (1994: 25) will say that the return to the community in the new social conditions and with a Welfare State in crisis, means to retake a protagonism that seemed forgotten or submerged, revitalizing "its will to count, to have a role in the social processes, in the decision making, in a word its will to participate. The demand for participation resurfaces from the social strata and spheres from which it had been expelled in the belief, somehow shared or assumed by too many subjects, of its uselessness".

- To affirm each person's protagonism as a subject and agent of the processes of social change, from his or her immediate environment and with the perspective of an increasingly interdependent and globalized society. The 'subject' of development, even if it is described as sustainable, is not the environment. Because, obviously, development refers to people and communities and not to objects, with all the consequences that this entails: "it is a matter of involving each subject in the defense of its natural and cultural environment, contributing both to the promotion of identities and to the redefinition of local autonomies. A mission that must be articulated from the biography that each person contributes to the common history, contextualizing it in the social spaces and times that are proper to him, from a strict respect for human rights and the inalienable aspiration to progressively improve the quality of life" (Caride, 1997: 225).

We cannot ignore, as Leff (1986: 187-195) emphasizes, that the environmental principles of development are based on a critique of the homogenization of productive and cultural patterns, vindicating the values of cultural plurality and the preservation of the ethnic identities of peoples. This reading has been repeatedly contravened by the dominant paradigms of the market economy. This is why it is necessary to define the environment also as an ethical and political stay: "as a condition for the implementation of community management projects of natural resources at the local level and as an effective means to achieve the objectives of sustainable development.... Nature ceases to be just an economic resource and is transformed into a cultural heritage; multiple resource management strategies offer principles for optimizing the sustained supply of resources while preserving the conditions of production sustainability, based on a differentiated appropriation of satisfiers in time and space, as well as on a more equitable distribution of resources and wealth". All this must lead to a new economic order based on local environmental management, in which the aim is to provide local populations with the minimum support and means necessary for them to develop their own self-management potential in practices

As we have argued in other writings (Caride and Meira, 1998, 2004; Meira, 2001), the growth of the global market and the establishment of its mechanisms of political, labor, environmental, etc., interdependence tends to break down smaller-scale economies to the point of threatening the very foundations of human existence and, in the long term, of the biosphere itself. tends to break down smaller scale economies, to the extent of threatening the very foundations of human existence and, in the long term, of the biosphere itself, it is understandable that for many "non-Western" communities sustainability means fundamentally a way of resisting progress; or, as Sauvé (1999) explains, translating the endogenous perspective into an alternative form of development, that it
The community and local dimensions of the project imply a way to address cultural disintegration and the disintegration of small economies.

In other words, it is not only a matter of taking the initiative or undertaking transformations, but of doing so with the greatest possible coherence between local and global processes, always in accordance with a vision of the future that is both humane and ecologically desirable.

Apart from other considerations, in which different processes that take local communities as spheres of explanation and construction of complex social realities (methodological strategies oriented to knowledge and social intervention, models of community action and intervention, etc.) are specified, which are diversified in denominations that confer certain substantivity to community work (community studies, community organization, community development, community promotion, community social work, etc.).), in relation to Environmental Education it is worth mentioning the existence of the so-called *Community Environmental Education Programs* (PEAC), defined as "those educational activities developed within the framework of a small community -neighborhood or village- and oriented to the attainment of knowledge and activities in relation to some environmental problem of the community itself (fires, pollution, waste, water management, etc.) (Sureda and Colom, 1989: 226). The PEACs may or may not be part of broader programs (Local Agenda 21, Integral Development Programs, etc.), although they are normally conceived as educational tools within the framework of broader development initiatives.

Most of these programs, which begin by combating the functional illiteracy that exists in environmental matters in many communities and social sectors, have the following objectives, among others:

- Increase the participation of communities in the task of rebuilding a healthy environment and achieving a better quality of life;

- Promote the constitution of community groups in the local power spheres, for the defense and conservation of the environment, and at the same time to transversally address other social purposes: employment generation, promotion of public health, community self-organization, democratization in decision making, etc..;

- Encourage the knowledge and research of the communities about their own environmental problems, oriented towards awareness and self-management;

- Strengthen public commitment and the sense of personal and collective responsibility in decision making and in the acceptance of the consequences of all types of measures that generate environmental impact;

- Strengthen the identity of human groups involved in migratory processes, mainly those deriving from the exodus from the countryside to the cities or from the socioeconomic south to the north, or whose reality is altered by external impacts that question or significantly transform their existence;

- Generate greater solidarity and cooperation among communities and social groups that share the same territory, encouraging their participation in the elaboration and management of local development projects.

In the Anglo-Saxon literature of the last decade on the field of AE, two concepts have been widely used to identify the principles of this participatory and community-based approach: *empowering* and *ownership*. The first is usually translated literally into English as "empowering", and its originality is questionable to say the least, since it refers us to the proposals of the Popular Education Movement that spread throughout Latin America in the 1960s and 1970s, with such relevant and influential figures in the subsequent development of Critical Pedagogy as Paulo Freire.

Empowerment" consists of offering the tools (educational, political, social, cultural, etc.) to a community so that it becomes aware of its reality, organizes itself and "assumes the power" to transform it.

The second, *ownership*, can be translated as "appropriation" and refers us to the same theoretical field: what, if not, was the meaning in Freirean pedagogy of the concept of conscientization rather than a critical and conscious appropriation of the world as a premise to be able to act in and on it in order to transform it. These are, in short, new "concepts" coated with Anglo-Saxon cultural hegemony, which do nothing more than update and invigorate old educational paradigms. The interactive encounter and the interacting attitude, which for Gudynas and Evia (1993: 97) are basic components in the initiative undertaken by social ecologists at the community level, should represent priority objectives for educators; an interaction that translates into: intentionality of sharing and understanding the affective, ethical, cognitive and political components of reality; the attitude of respect towards the people with whom they interact; the attitude of attentive observation of what happens in the field of praxis; the attitude of dedication to their praxis; and the critical and reflective attitude, which does not seek to fall into the easy militancy that forgives the superficiality of the work, or that is sheltered in dogmatisms.

The environmental or social ecological educator - in the vocabulary of Guydinas and Evia - must pay attention to various aspects of what surrounds him, so he must understand the natural and socially constructed framework of the environment where his praxis develops. In this sense, the differential identities of the environment in each community and for each community must be an inevitable reference for Environmental Education.

The discourse constructed up to this point forces, as a conclusion, the need for Environmental Education developed in community contexts to abandon its categorization as a "non-formal education", lacking systematics or entity *by* and *in* itself. On the contrary, it claims to be "formal" and "significantly" constituted in social scenarios and pedagogical practices, with objectives and methods, techniques and strategies, contents and actors, experiences and practices, etc. that cannot be interpreted either as a way of making viable some kind of educational denial (in this case, that represented by "formal" or school education) or as the expression of a parallel, subsidiary or partial action of education.

Hence, we claim (Caride and Meira, 2004) the denomination of Community Environmental Education as a way of recognizing and delimiting the profiles of a pedagogical and social practice that makes its own the commitments to advance towards a sustainable society, at least as long as words continue to exercise some kind of symbolic and/or material power.

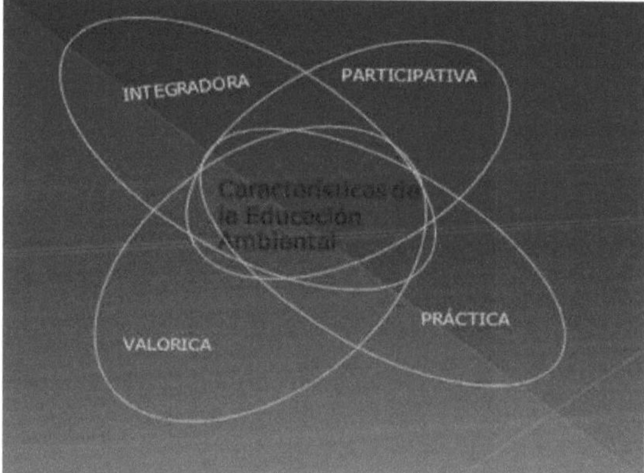

ENVIRONMENTAL EDUCATION BASED ON THREE APPROACHES: COMMUNITY.

Man, since his emergence, has enjoyed the riches of nature, but at that time he did not affect them to a considerable degree. In the later stages of the development of human society, man's capacity to modify his environment grew, and consequently, the need arose to safeguard nature from the harmful effects of this activity.

It was at the beginning of the second half of the 20th century when unprecedented problems began to be perceived that affected the very survival of the human species on earth.

Our society is called to train citizens capable of playing their role in favor of the planet and its socio-natural environment, based on solid knowledge, principles and convictions, ensuring the relational balance of the plurality of biotic, abiotic and social factors that make up our living space.

Education as a process and the school as an institution play an essential role in this battle, since they must involve all members of society in the search for solutions to solve environmental problems, providing them with the knowledge, skills and motivations necessary for an adequate interpretation of the world and social action consistent with their needs and demands.

To ensure that education achieves this essential purpose, the formative introduction of the environmental dimension in its socio-cultural integrality is required. This reality implies a treatment of environmental issues in a coherent and meaningful way, which will allow students' cognitive activity to be in constant development in order to integrate knowledge.

Since its conception in Stockholm, Environmental Education has been a permanent educational process aimed at making individuals and society in general aware of their environment and acquiring knowledge, skills and values that allow them to develop a positive role, both individually and collectively, towards the protection of the environment and the improvement of the quality of human life.

It is evident then, that Environmental Education does not present barriers of age or educational system, so that at any time the individual is capable of positively orienting his impressions and values regarding the environment.

This conceptualization leads to an analysis of the existing positions on how to introduce environmental issues in the Educational Teaching Process and the corresponding assumption of criteria for the basis of the proposal for its introduction.

We conceive the **environmental dimension** as: an approach that in an educational and research process, is expressed by the systemic character of a set of elements that have a determined environmental orientation: expressed through the links between environment and development; which are consequently interconnected, where the functions or behaviors of one act and can modify those of the others.

In other words, the environmental dimension in an educational process has a systemic, interdisciplinary and community character, which will be referred to as approaches later in this research.

In the case of the environmental dimension of a study plan, for example, its introduction will consist of the incorporation of an integrating conception of knowledge, habits, skills, attitudes and values, consciously designed and contextualized, that crosses the entire study plan, so that it is established that each of these contents contributed in the educational teaching process and that results in an integral formation of the subject, which is expressed in his/her actions towards the environment, with respect to his/her surroundings and environmental problems.

Therefore, it should be taken into account that the introduction of the environmental dimension in the educational system will lead to changes in the theory and methodology of the curriculum, which will enhance critical appraisal, modification of attitudes, values and the development of responsible behavior towards the environment and its surroundings in particular.

Environmental Education, besides being a permanent learning process, where values are affirmed, is a process aimed at improving the quality of life and the conditions of the population, human relations, its culture and environment, recognizing it as an educational resource, protecting the environment and understanding the relationship between man, nature and society.

At the national and international level, Environmental Education has been developed from the interdisciplinary, multidisciplinary and transdisciplinary approaches, in addition to the community approach. Through this article we intend to contribute to the formation of Environmental Education through the integration of community, systemic and interdisciplinary approaches.

The **community approach** is an approach that has been worked on a lot in recent times, due to the need to influence our students with the objective of forming environmental attitudes and values to appease the crisis and transform the predatory attitude of man, due to the ecological problems existing on our planet.

Environmental education should develop in students the capacity for critical observation, understanding and responsibility towards the environment, which is characterized by its multivariety. A fundamental principle of this is the contextualization of the content to the environment where the student lives, which is why it is par excellence communitarian, since the community is its fundamental field and its problems must be part of the content of the activities.

The problems and their causes should be studied and analyzed from the local to the global with a progression of connected continuity: micro, macro and vice versa. Starting from the solution of problems close to the life of the center or community, that is to say, to place the students in front of local environmental realities and from them, they can go into other regional or global realities.

A mistake that is made in the treatment of this approach is to give importance to the environmental phenomena and problems that occur on an international scale, not that we do not take into account these problems that occur in the world, because it is necessary to know them, because they affect our planet and also affect us, but we must take into account the problems that occur in our country, our province, in the municipality and more specifically in the environment of the school, the community where we live, to know the causes that provoke them and their possible solutions. If we want to achieve an awareness of the main problems of the community, we have to carry out activities with the students that allow them to identify these problems, analyze the causes of their emergence, the consequences for the life of the community and their involvement in the practical solution of them, an issue that is very much taken into account in the objectives of education for all levels of education in the country and in this research.

At present, there is no discussion on whether the environment is much more than nature or natural ecosystem, that is, it is a complex system intimately related where different concatenated elements such as man, nature, society, social and cultural relations, etc., are taken into account.

It is necessary to understand the responsibility that should be assumed towards the environment and why it is necessary to favor Environmental Education of students in the institutional and social context as a more general space for action, which is why this community approach is necessary. Working this approach in the Educational Teaching Process implies the systemic and systematic integration of Environmental Education, from an "Environment-School-Community" linkage perspective, hence the importance of contextualizing the environment where the school is located, of inserting through the educational system environmental contents to form in our students an environmental culture, in order to achieve correct behaviors towards the environment.

Therefore, it is necessary to work on the **systemic approach** in Environmental Education.

All environmental problems necessarily have a systemic constitution, considering them as an organized whole, composed of parts that interact with each other. Therefore, understanding the environment as a system in which the elements that integrate it are interrelated is a fundamental characteristic of the environmental dimension. The environment also manifests a **systemic vision**, where the components of the system are integrated in the physical, biotic, economic and sociocultural environment.

As can be seen, the fundamental characteristic of the systemic approach is not so much in the composition of the elements that make up its parts, but in how these parts integrate with each other to form a dialectical unity (so that the change that occurs in any of its elements affects the others) and how the integration between them leads to development.

From this analysis, it can be stated that a Teaching - Learning process based on a systemic vision, should be characterized by:

The integration of its constituent elements.

The reciprocal enrichment of the related subjects.

A holistic conception of reality.

The transformation of traditional teaching and learning styles to produce changes from the didactic point of view, which necessarily leads to the formulation of projects, programs and educational strategies that respond to the real needs.

Pain related to has been addressed in different research.this terminology or theory. We understand the system as: "a set of elements that relate to each other and to the environment". The scientific acceptance of this approach gives epistemological and methodological operability to the **systemic approach**, also understood as: "a paradigm of character, insofar as it represents a complete constellation of beliefs, values, techniques and world view, shared by some members of a certain community".

The systemic approach as a scientific conception evidences an instrumental duality of unobjectionable value: "as a descriptive analysis methodology and as a system optimization strategy". For this reason, before defining the environment, it is necessary to analyzo from a philosophical point of view the definition of **system:** "set of interrelated elements that constitute a certain integral formation". These are the reasons why the **systemic approach** supports the theoretical basis for the treatment of **environmental** problems, which has gone through different moments in its historical evolution.

The systemic approach appears as a methodological tool whose purpose is to identify in a coherent framework the set of factors, states and interactions that characterize the appearance of a phenomenon of the existence of any environmental problem.

The systemic approach in Environmental Education requires a project approached from a global vision that considers that it is an open system in which the whole is more than the sum of its parts, in which the knowledge of the interrelationships is more explanatory, where the interdisciplinary treatment is sought, the structure and functioning are valued, taking into account dynamic and evolutionary aspects and the realization of the system given its complexity. The great challenge of Environmental Education is to know how to grasp the totality in flowing movement, which supposes a teaching-learning model in which additional and juxtaposed knowledge is not proposed, but rather it is necessary to establish connections and relationshipsof knowledge in an undivided and permanently changing totality.change. This integrative treatment of knowledge requires an interdisciplinary approach.

Interdisciplinary approach.
Interdisciplinarity representsa set of interrelated disciplines with defined relationships, so that their activities do not take place in an isolated, dispersed and fractioned manner. It is born with the individual character of diverse subjects that make evident their interdependencies and with them it is possible to give a global and less schematic vision of the problems. That is to say, the articulation of the different disciplines in order to understand a process in its totality, in order to move on to the analysis and solution of a particular problem.

The incorporation of this interdisciplinary approach to educational practice should be carried out gradually, which presupposes the realization of pedagogical collectives, years and disciplines, in order to achieve an adequate organization of teaching, which contributes to the students' understanding of the complex structure of the environment, as it results from the interaction of its physical, biological, social and cultural aspects, as well as to provide a clear awareness of the political, economic and ecological interdependence of the world.

The aim is therefore to raise awareness of the problems that pose obstacles to individual and collective well-being, investigate their causes and identify ways to solve them. In this way, they will be able to participate in a collective definition of strategies to solve the problems that affect the quality of the environment.

Interdisciplinarity is understood as ".... methodology that characterizes a teaching, research or management process, in which an interrelation of coordination and effective cooperation between disciplines is established, but maintaining their theoretical-methodological frameworks...", a concept to which the researcher subscribes, due to its great connotation from the methodological point of view to implement in the pedagogical practice.

Interdisciplinary integration processes involve a more organic relationship between subjects, where each subject contributes conceptual schemes, integration methods and ways of analyzing problems through close and coordinated cooperation.

The dialectical-materialist philosophical principle of concatenation of phenomena is reflected in teaching through the contents common to several subjects, in the Teaching and Learning Process.

The progress of knowledge is achieved in the movement of thought, which passes from less deep and general links to establish deeper and more specific links between the facts, processes, phenomena of this infinite world.

Interdisciplinarity is not only an epistemological criterion, an instrumental and operative system, but also a way of being. It expresses the multiple character of relationships and the orientation of meaning according to the orders it establishes.

The interdisciplinary nature of Environmental Education through this research will be worked from how to insert through the educational system environmental contents to form in our students an environmental culture and achieve correct behaviors towards the environment, it is not only to know it, that is to say, it is not enough to educate for nature using it as an educational resource, but:

Environmental education: where environmental issues are addressed in the classroom or workshop (especially in natural and urban environments).

Education **in** the environment: a study of the environment in which the school is located, of the environment surrounding the students in which they develop, all this from a naturalistic point of view.

Education **for** the environment: it leads to an action aimed at changing attitudes, to the formation of values, to preserve the natural and/or urban environment, to influence it in a spirit of preserving it if it is not damaged or transforming it if it is.

Based on the above, it can be stated that the school as an educational institution is in charge of forming an integral personality, capable of promoting a sustainable development, through a planned, organized and coherent pedagogical process.

This requires a teacher with a high level of integrity, who guarantees, in addition to the necessary knowledge, the development of skills and the formation of values that our society demands today for the care and conservation of our environment, a teacher who educates our students environmentally. Hence the need to include in the curriculum the environmental dimension with a holistic, developmental perspective, as established in the principles of Environmental Education towards sustainable societies.

Environmental Education should be materialized taking into account the system of educational influences, where the core is the school, where the teacher, through the proposed methodological indications, achieves the development of critical thinking and with this the student feels responsible for the environment of which he/she is a part.

The development of an interdisciplinary didactic model where the approaches (systemic, community and interdisciplinary) of Environmental Education are interrelated makes possible a greater orientation, teacher-student interaction and articulation between knowledge and environmental attitudes.

The students, being the protagonists of the process, identify the environmental problems related to the content under study, make evaluations, analyses and come to propose actions, that is, they go through different stages where they manifest their transforming actions before the environment.

Taking into account the considerations that should be taken from Environmental Education for this research, there is a need to achieve in the students a learning process that provokes a change in their behavior and in the attainment of essential values. The inclusion of the environment in the programs of the exact sciences subjects in pre-university education leads to significant changes in the educational system, from its purposes to the contents and methodologies of its teaching, so as to redefine the type of person to be educated, according to the future scenarios of their performance.

HISTORICAL PROCESS OF ENVIRONMENTAL EDUCATION

Since 1972, the year in which the Declaration on the Human Environment was signed at the Stockholm meeting, there have been concrete expressions of and commitments to sustainable development in the world, although concern about the unsustainable management of our planet predates this date. But this declaration, with its subsequent repercussions, is the one that marked a fundamental milestone in the advance towards the understanding of the urgent need for a change in development processes.

The Declaration stated that "man has the fundamental right to freedom, equality and the enjoyment of adequate conditions of life, in an environment of a quality that permits a life of dignity and well-being, and has the solemn obligation to protect the environment for present and future generations".

Key institutions such as the United Nations Environment Program (UNEP) emerged and in 1975 it was proposed that UNESCO take charge of implementing the Interdisciplinary Environmental Education Program (IEEP), as fundamental contributions to achieving a change in the vision of development and in the education that could lead to it.

The Belgrade Meeting on Environmental Education was held in 1975, and promoted an international effort by UNEP and UNESCO to better understand and put into practice this new education, which was later outlined in greater detail and with extraordinary vision at the Intergovernmental Conference on Environmental Education, held in Tbilisi in October 1977. Between Belgrade and Tbilisi, regional preparatory meetings were held, such as those for Africa (Brazzaville, 1976), Latin America and the Caribbean (Bogotá, 1976) and Europe (Helsinki, 1977), where important debates were held and the vision of the new type of education for the future was broadened. Some of these interesting contributions are expressed in statements such as the following:

"Environmental Education should encourage the establishment of a value system that is in harmony with the traditional cultural environment.... Aggressions, conflicts and wars produce disastrous effects on man and the environment. Therefore, education should promote peace and justice among nations" (Brazzaville, 1976).5

"Environmental Education should aim to strengthen the axiological sense, contribute to the collective welfare, be concerned with the survival of humanity" (Helsinki, 1977)6.

This global vision, related to values, to peace and justice, to collective well-being, opened up a conceptual dimension that was later strengthened at the Tbilisi Meeting, whose final Declaration emphasizes:

"Environmental Education is in fact education as it should be understood and practiced in our time. Environmental Education, besides being oriented towards the community, should involve the individual in an active process aimed at solving problems arising in the context of specific realities, encouraging initiative, responsibility and a prospective sense of a better tomorrow" (Tbilisi, 1977).

Latin America was undoubtedly one of the regions of the world that enthusiastically embraced the commitment to environmental education. Changes were made to curricula by the Ministries of Education, pilot projects were carried out and evaluated, and various actions were proposed to insert the environmental dimension into the educational curriculum.

In 1987, the report Our Common Future was published by the Brundtland Commission, created three years earlier, with the participation of experts from various regions of the world, some of them from Latin America. This report presented the definition of sustainable development that is widely used today, and which served

as a reference for the documents of the 1992 Rio Summit:
"Sustainable development is that development that makes it possible to meet the needs of the present without compromising the ability of future generations to meet their own needs."
Numerous heads of state and government welcomed the Rio Declaration, signed the Conventions on Biological Diversity and Climate Change, and endorsed the Action Program known as Agenda 21.
Chapter 36 of Agenda 21 is devoted to education, public awareness and capacity building, proposing the reorientation of education towards sustainable development (within the framework of the recommendations of the World Conference on Education in Jomtien, Japan in 2000) and the reorientation of education towards sustainable development (within the framework of the recommendations of the World Conference on Education in Jomtien, Japan in 2000).
1990), and proposing a series of objectives and activities to achieve them.
The Convention on Biological Diversity, for its part, dedicates its Article 13 to Education and Public Awareness, proposing the promotion and fostering of understanding of the importance of the conservation and sustainable use of biological diversity.

Parallel to the formal meeting convened by the United Nations in Rio de Janeiro, the Global Citizen Forum was held, with the participation of thousands of people and independent institutions from around the world, where a Treaty on Environmental Education towards Sustainable Societies and Global Responsibility8 was proposed, consisting of a series of axiological, political and methodological principles to generate values, attitudes and behaviors in line with the construction of a sustainable, fair and ecologically balanced society.
After Rio, the international context was enriched by a series of new meetings and commitments on important aspects related to development: the Cairo Conference on Population, held in 1994; the Conference on Social Development, held in Copenhagen in 1995; the Conference on Women, held in Beijing in the same year; the Istanbul Conference on Human Settlements in 1996, etc.
Ten years after Rio, in 2002, the World Summit on Sustainable Development (WSSD) was held at the
Sustainable Development, also convened by the United Nations.
The final document of this meeting or Action Plan has only brief references to education in general and does not dedicate a special section to environmental education9.
However, important regional and global meetings are being prepared for 2003, such as the First World Congress on Environmental Education, to be held in Espinho, Portugal, in May, and the IV Ibero-American Congress on Environmental Education, to be held in Havana, Cuba, in June.

DEFINITION, SUBDIVISIONS, OBJECTIVES AND CHARACTERISTICS OF ENVIRONMENTAL EDUCATION

DEFINITION OF ENVIRONMENTAL EDUCATION

"Environmental education is a formative process through which the individual and the community are made aware of and understand the forms of interaction between society and nature, their causes and consequences, so that they can act in an integrated and rational way with their environment "10.
In this definition, it is important to note that the educational process not only seeks to increase the knowledge of the target population, but also to understand the fundamental interactions between humans and nature, all with a specific purpose: action.

In other words, Environmental Education is proposed as an integral and systemic activity, with two central emphases: analysis, knowledge and understanding of interactions and participatory social action towards environmental improvement.

SUBDIVISIONS OF ENVIRONMENTAL EDUCATION

Environmental Education has traditionally been divided into formal Environmental Education, non-formal Environmental Education and informal Environmental Education.
Formal Environmental Education is that which is carried out within the framework of formal educational processes, that is, those leading to certifications or degrees, from preschool, through primary and secondary school, to university and postgraduate education. The forms of expression of this education range from the incorporation of the environmental dimension in a cross-cutting manner in the curriculum, to the insertion of new related subjects, or the establishment of school educational projects.
Non-formal environmental education is aimed at all sectors of the community, in order to provide greater knowledge and understanding of global and local environmental realities, so as to promote improvement processes that incorporate the various groups of society, men and women, ethnic groups, organized communities, productive sectors, government officials, etc. It is generally expressed in the realization of workshops, seminars, courses and other formative activities, inserted in community social development programs, or in educational plans of public or private organizations, at national, regional or local level.
Informal Environmental Education is oriented in a broad and open way to the community, to the general public, proposing guidelines for individual and collective behavior on the alternatives for an appropriate environmental management, or raising critical opinions on the existing environmental situation, through different means and mechanisms of communication.
Examples are radio or television programs, educational campaigns, articles or offprints in the written press, the use of flyers, the presentation of theatrical plays, the staging of musical performances, etc.

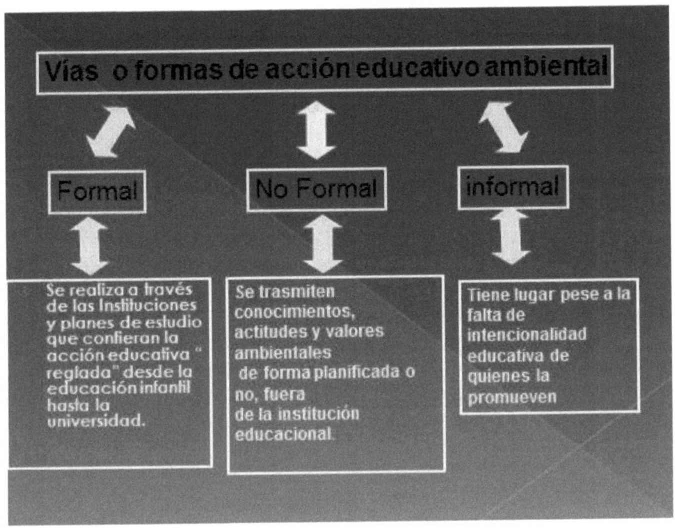

OBJECTIVES OF ENVIRONMENTAL EDUCATION

The Belgrade Charter on Environmental Education, produced and adopted at the end of the Belgrade Conference convened by the United Nations Educational, Scientific and Cultural Organization (UNESCO) in October 1975, includes the following objectives of Environmental Education:

Environmental Education and Biodiversity Conservation in Community Development

Project "Biodiversity Conservation and Sustainable Management of the Huasco Salt Flat".

a. To educate and awaken environmental awareness.
b. Generate knowledge in individuals and social groups to gain a basic understanding of the environment as a whole.
c. To develop attitudes in individuals and social groups, based on the acquisition of social values and interest in the environment.
d. To discover and cultivate people's abilities to solve environmental problems, by themselves and/or acting collectively.
e. Stimulate participation, helping individuals and social groups to deepen their sense of responsibility and to express it through decisive action.
f. To develop the capacity of evaluation in individuals and social groups, in order to evaluate Environmental Education measures and programs.

This statement of objectives, although at the time it was an important step in the process of organizing and systematizing environmental work from an educational point of view, should be enriched and updated, incorporating, among other elements, a holistic and integrating conception of the environment, as well as the futuristic forecast of sustainable development and its impact on the quality of life of the people.

Taking the definition of Environmental Education as a conceptual basis, the following objectives, among others, could be added:

g. Empower the population, individually and collectively, to assume in a participatory manner the environmental management of the geographic space it occupies.
h. Contribute to the construction of an integral and holistic vision of the environment, providing intellectual tools and means to access and build environmental knowledge.
i. To promote and stimulate actions aimed at achieving sustainable levels of development on a human scale, providing conceptual and instrumental bases to improve and maintain optimal quality of life conditions for all.

Current environmental education is conceived in close relation to the dynamic conception of the environment, and has stronger links with environmental management than with the simple description of environmental problems. This fact marks one of its central characteristics: the link with sustainable development and participation.

CHARACTERISTICS OF ENVIRONMENTAL EDUCATION

On this premise, the characteristics of Environmental Education, which were proposed at the Intergovernmental Conference on Environmental Education held in Tbilisi, Georgia, in 1977 and ratified over time, can be presented in general terms as follows:

a. Globality and integrality. In other words, the environment is considered in its totality with a holistic and integrating approach, examining the natural and social aspects in interaction.

b. Continuity and permanence. It must be an uninterrupted process that occurs and accompanies human beings and social groups in all stages of life.
c. Interdisciplinarity and transdisciplinarity. Its conceptual and action field encompasses and transcends the artificial limits of the different disciplines of human knowledge.

d. Spatial coverage. Its influence covers the local, regional, national and international levels: it must be situated in specific situations as well as in their near and distant contexts.

e. Temporality and sustainability. It models the management of the current situation and the vision of the future; that is, it focuses on today's environmental situations and those that may arise, within a historical perspective, towards the construction of desirable and possible alternative futures for life in all its forms of manifestation.

f. Participation and commitment. Engages and stimulates participation from different sectors of the population in the achievement of sound environmental management through local, regional, national and international cooperation.

g. Foundation for development. In this sense, it uses diverse methods to facilitate knowledge and understanding of environmental situations, deepening in those methods that make participatory processes viable; it influences and guides development plans, strategies and action methods to achieve sustainable development on a human scale.

h. Linkage with reality. Its action is aimed at achieving a close and active link with the local, national, regional and global reality.

i. Universality. By its conception and orientation, it is directed to all sectors of the population, to all age, ethnic and gender groups, and to all educational and social levels in order to involve them actively, towards a participative environmental management.

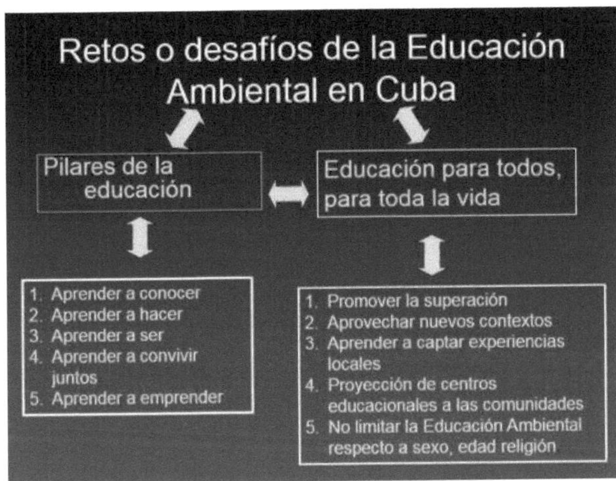

ELEMENTS OF PARTICIPATORY ACTION RESEARCH IN ENVIRONMENTAL EDUCATION

STAGES OF PARTICIPATORY RESEARCH
Participatory research can be carried out in approximately eight stages of work, not necessarily sequential, as some of them are carried out in parallel, according to the purposes and needs of the process.

The stages may be as follows:

Stage 1: Questioning and exploration

In it, the global situation is identified, the characteristics of areas and populations are determined, the main questions and interrelationships are addressed, and a rationale for the work is established.

Stage 2. Topics to investigate
At this stage, the topics to be investigated or topics of analysis are presented, the knowledge to be obtained is discussed, as well as the disciplinary and interdisciplinary approach of the work.

Stage 3. Thematic location

Placement in a broader problematic framework: analysis of the global situation, desirable scenarios, problems to be solved, basic planning, etc.

Stage 4. Theoretical-practical orientation

Some starting assumptions are established, the principles of the participatory research to be carried out, as well as the characteristics of the emerging design, the approach to achieve the objectives, the guidelines for the search for information and the use of techniques, as well as the elaboration of the support instruments.

Stage 5. Participating research group

A group of participating researchers is created and set up. Consultation workshops are held with counterpart institutions or entities and training workshops for those who will lead the work directly with the community (qualified as promoters).

Stage 6. Definition of the Study Universe

The geographic field is delimited, as well as the participating communities and individuals. The possibilities of access to the communities under study are analyzed and the qualitative and quantitative representativeness of the sample is established.

Stage 7. Data collection and systematization

The instruments for the application of the various techniques are designed and prepared, and training workshops are held for the use of the instruments. The techniques are applied and the data is collected and organized. The main results are systematized.

Stage 8. Interpretative analysis of results and conclusions

The systematized results are analyzed in order to explain them and reach the main conclusions. They are ordered and studied, and the respective reports are prepared. The methodology of participatory research includes the application of several techniques11 , as well as their mutual articulation, the differentiated examination of the responses and also the search for related and contrasting elements that make it possible to distinguish attitudes and opinions according to their situation in same-gender or mixed groups, and in relation to the age ranges studied.

PARTICIPATORY RESEARCH AND ENVIRONMENTAL EDUCATION

Environmental education should include participatory research processes, of varying depth and level, in accordance with the objectives of educational actions. In turn, participatory research on environmental issues should become an educational tool.

The process of studying complex environmental situations leads to formative situations that can be used positively. What is fundamental is to keep in mind that environmental education is not only about the simple reproduction of pre-existing knowledge, adequate or summarized to be understood by large social groups, but also includes research and the creation of new knowledge.

In accordance with the characteristics of environmental education, research must be participatory and produce concrete results for the improvement of environmental conditions and the quality of life of the population. In addition, this research proposes processes of approaching realities in which we all can and should contribute.

Modules V and VI present more details on the techniques and instruments that can be applied in environmental education processes, with a research and participatory approach.

PARTICIPATORY DESIGN OF COMMUNITY ENVIRONMENTAL EDUCATION PROJECTS

Community environmental education is part of non-formal environmental education, and corresponds to educational work on the environment and sustainable development aimed at urban and rural communities, in order to train them and encourage action to improve their environmental conditions and strengthen the processes of conservation of biological and cultural diversity.

A community environmental education plan is also part of a participatory research process, insofar as it includes various moments of analysis of the realities in which the population lives.

There are several ways of approaching the planning of an Environmental Education Plan, linked to the implementation of specific community environmental recovery actions. Here we present a simple scheme, considered "classic", which is useful in different contexts.

When attempting its implementation, each step must be examined analytically and critically, in order to be able to adjust the tasks conveniently to the real requirements of the community and the region under consideration.

The possible steps to develop and implement such a plan are as follows:

STEP 1: Analysis of the environmental and biodiversity conservation situation. Establishment of the main environmental problems suffered by the community and determination of those that will be considered a priority for the plan. This should be done in a participatory manner, using techniques that are considered appropriate for the type of community where the work is being carried out (see techniques and instruments in modules V and VI). The prioritization of problems should also be carried out in a participatory manner, using dynamic selection methods and considering technical alternatives for the weighting of each problem (criteria of greatest coverage, incidence, affected population, etc.).

STEP 2: Definition of alternative solutions to the problems identified. Once the problems have been selected, each one of them needs to be analyzed from the perspective of its possible technical, organizational or behavioral solutions, with the intention of defining possible additional studies to be carried out, or adjustments to be incorporated into the process.

STEP 3: Analysis of the community's perceptions of the problems and the degree of knowledge they have about them. How the problem is expressed in them and how much they know about it, intuitively or explicitly.

STEP 4: Elaboration of central guidelines for the solution of the prioritized environmental problems and detection of the educational topics to which each of them is related. The environmental problems will be prioritized according to educational action guidelines to be defined in this step. To this end, it is important to bear in mind that, to a greater or lesser extent, there are environmental management processes underway within the same community or at the municipal level, and that these guidelines, as well as the entire plan, should be closely linked to this work and to previous intersectoral experiences.

STEP 5: Establishment of general purposes and coverage. The purposes may be associated with general educational aspects for the population, environmental awareness, group and individual participation, collective organization, or they may be directed to achievements related to levels of solution of specific problems where there is a need to educate the population. Coverage will indicate the scope of the work, in terms of geographic area and the people and groups to which it will be directed.

STEP 6: Formulation of objectives and development of the environmental education action plan. The community should contribute in a participatory manner to the formulation of specific objectives and the action plan, which will also include the activities to be carried out for each objective and the expected results of each activity. A general strategy will also be defined, including the sectors involved and the responsibilities of each activity.

STEP 7: Implementation of the action plan. The active participation of the community is of vital importance, and must be present in the various activities, according to the capabilities, knowledge and availability of individuals and groups.

STEP 8: Elaboration of informative messages according to the thematic priorities, for different media. The messages will be elaborated based on what was obtained in the previous steps, choosing the most appropriate forms of communication for the community and its cultural characteristics, for the treatment of the selected problems and for the geographic environment.

STEP 9: Selection of strategies to deliver these messages (forms of dissemination, reinforcements, links with specific tasks, receptivity balance, etc.) These include the means of communication to be used, the forms of dissemination of the messages, the subsequent reinforcements, the links between the messages and specific tasks to be performed or underway, the gradual analysis of receptivity, etc.

STEP 10: Monitoring and evaluation of the first stages of the plan and identification of modifications to be made to the plan. It is necessary to establish a concrete monitoring and evaluation plan for the entire Plan. These results should be part of a feedback process for all steps.

STEP 11: Follow-up of results and alternatives to achieve sustainability. It is important to establish concrete lines of follow-up of the process, articulated with the evaluation and fulfillment of each of the activities and the expected results.

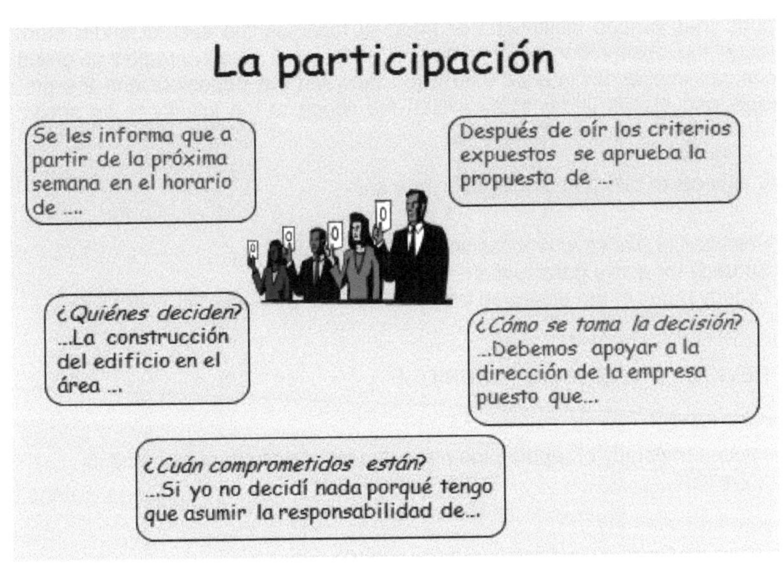

COMMUNITY PARTICIPATORY TECHNIQUES IN ENVIRONMENTAL EDUCATION

PARTICIPANT OBSERVATION TECHNIQUES

The objective of this technique in environmental education is to gather both qualitative and quantitative information on natural and social characteristics in a given place and time, and on general environmental problems, through documents and reports that various institutions or persons have on the subject under study, and through the observation of ongoing processes in a locality, region or group. This technique implies an intense interaction between the educators and the groups of people and places under study, within the scope of the activity to be analyzed for environmental education purposes.

The aspects to consider in this technique are:

- Selection of places and occasions for observation
- Strategy for active participation
- Orderly notes of the observed facts: Guide for the notes.

. PREVIOUS DIAGNOSTIC TECHNIQUE

It includes the study of reports, the procurement of documentation and its arrangement.

. IN THE SOCIAL, ECONOMIC AND CULTURAL FIELDS

Preliminary diagnoses require the analysis of several issues recognized as determinants in the social, economic and cultural fields:

- Socio-economic and cultural characteristics of the communities
- Migration and miscegenation process
- Gender relations in communities (history and evolution)
- Community relations with local NGOs (projects, characteristics)

The following are some subtopics that can be included:

- Total population of each community disaggregated by sex
- Gender hierarchies in the rural world
- Main myths and legends in the selected communities
-   Basic services and personnel working in them (health, education, etc.)

IN THE ENVIRONMENTAL FIELD

Some of the issues to consider may include:

- Natural characteristics of the areas inhabited by the communities
- Maps of the study areas
- Main elements of the Biodiversity of the zones
- Natural resource management
- Major environmental and biodiversity conservation issues

The following are some subtopics that can be included:

- Altitude of each community (meters above sea level)
- Surface area of each of them
- Climatic characteristics
- Land types and land use
- River basin to which each community belongs
- Representative ecosystems
- Representative species of flora and vegetation
- Representative fauna species
- How people manage the resources and the environment where they live (methods, techniques, infrastructure, etc.).
- Ways of using resources to ensure their survival
- Main advantages and threats to biodiversity in the study areas

SEMI-STRUCTURED DIALOGUES AND INTERVIEWS

This involves collecting general or particular information, based on a series of dialogues with people who are recognized for their specific knowledge (key informants), with groups of families, or with targeted groups (indigenous communities, community boards, teachers, etc.).

In contrast to interviews, dialogues only establish topics and seek an exchange of information and a fluid conversation between the parties. In the interview, in addition to the topics, specific questions and cross-examination are established.

The fact that the dialogue or interview is semi-structured indicates that it will not be governed by a closed questionnaire where the answers must be precise and brief, and that it will try to adapt to what people perceive and feel.

The steps to apply a dialog may be as follows:

- Define the criteria for selecting the people or groups with whom the dialogue will be conducted.
- Establish a guide of no more than 10 topics, in which you try to summarize what you want to research.
- Conduct dialogues.
- Analyze the results
- Comparison with other previous information

- Conclusions and fields of application.

The interview, likewise, proceeds in several steps:
- Selection of groups to be interviewed (by community, ethnicity, gender, representativeness and/or age range).
- Development of interview guides, including questions and cross-examination.
- Application of the interviews
- Ordering of answers
- Analysis of similarities, contributions, coincidences and contradictions
- Determination of results and conclusions

PARTICIPATORY WORKSHOPS TO DEVELOP ACTION PLANS

These are participatory work meetings, which should be held with a group of no more than 15 people, and can be carried out in three versions: one for women, one for men and a third mixed version for comparisons, balance sheets and final recommendations.
Workshops should always be conducted based on formats and in an orderly sequence to obtain concrete results. The results should be compiled and ordered in order to establish comparisons, conclusions and recommendations,

Possible workshop objectives

- Conduct a diagnosis of the main environmental and biodiversity conservation problems felt by the community, their priorities and solution options.
- Detect the main ideas and suggestions of the community (women and men, young people, adults and the elderly) as alternatives for achieving sustainable development and improving environmental conditions and biodiversity conservation.
- Establish community interests, ideas and options related to possible action plans.
- Identify the differences by gender in the topics worked on, as well as the coincidences found, as a basis for future strategies.

PARTICIPATORY MAPPING

A widely applied technique in community environmental education processes is the one that focuses on the participatory elaboration of maps. This technique makes it possible to visualize many important elements and become a reference for the analysis of diverse environmental situations, as well as for the proposal of alternatives to solve problems.

Some examples of maps that can be made jointly are:

MAP OF THE FARMS: Design of the land with what each one contains, crops, livestock, buildings, trees and shrubs, yards, domestic animals, indicating the distances between them and the people involved in the activities.
(men, women, boys, girls, landowners, tenants, etc.) and the places that present conflict situations.

MAP OF NATURAL AND CULTURAL RESOURCES IN THE REGION:
Design of the various natural and cultural resources, their location, state of conservation, access ways and routes, etc.

MAP OF THE HYDROGRAPHIC BASIN: Design of the basin, water sources, places of urban or rural concentration, communications, water pollution possibilities, uses of water sources, etc.

URBAN OR COMMUNITY NEIGHBORHOOD MAPPING: Design of the main streets, stores, housing, recreation centers, activities, environmental problems, energy sources, waste disposal centers, etc.

. SIMULATIONS OR ROLE-PLAYING

This technique is very useful for analyzing different environmental situations and the roles played by the various social actors in them. It is also possible to analyze alternative solutions and propose creative ideas to address the problem in a constructive manner.

The first step is to define the topic or problem to be addressed, as well as the characters involved in it, distributing among the participants each of the roles and establishing an initial agreement on what is to be worked on and the elements to be used as a starting point for its analysis.
Subsequently, the simulation work is started in an open manner, giving a time limit for it and keeping a balanced and dynamic rhythm of interventions.

The advantages of this technique include, among others, the fact that it is possible to

• Process personal and group information and expectations about what is shared in the social group on the subject (internal observation).
• Define relationships and communication between people and institutions (what is observed externally).
• Analyze the relationships that the participants have, their perceptions and contradictions regarding the topic.
• Work towards a final consensus and concrete proposals.

ENVIRONMENTAL GAMES

Numerous games with environmental contents can be carried out, open to the creativity of each social group and age group.

Among them, there are all the games derived from the ascertainment of our senses. For example:

SEE NATURE, SEE THE CITY: Explore carefully the visual field, divide it into four parts of 90 degrees each, and detail all the natural and social elements that are there, differentiating what we usually see and what goes unnoticed. Contrast with what different people have been able to see. Establish what is pleasing to the eye, and creates a beautiful landscape, and what disturbs it.

LISTENING TO NATURE, LISTENING TO THE CITY: Visiting various
places and detail the sounds, differentiating them as pleasing to the ear, disturbing,
etc.

Other environmental games are those created from board games already known to
most people. For example, mazes, monopoly games, card games, etc., in which the
traditional elements are replaced by themes related to society and nature, their
interactions and the problems and solutions derived from them.

FIELD TRIPS (OR FIELD TRIPS)

Field trips play a very important role in environmental education at all levels. This
technique requires several stages:

•   Choose the possible urban or rural geographic areas, because of their natural,
cultural or social interest, or their environmental problems, and outline the route, the
stations or observation points where the group will stop, the socialization activities to
be carried out there, the topics for reflection and analysis, the actual time for each
activity along the entire route, from the departure from the meeting point to the time
of return to the point of origin.

•   Provide prior information to the group of participants in the educational process,
providing basic information on the characteristics of the places to be visited, in order
to
that each person seeks more information and arrives motivated to go on the outing.

• Involve the participants in the preparatory phase of the field trip, so that there are
specific roles for each one of them, in terms of the operational part: food and
beverages to be brought, appropriate clothing, work tools (if applicable), etc. and the
academic part: details on biological or social aspects that can be studied in depth by
some of the assistants, etc.

• To follow some basic guidelines for the tour:

☐   contar con un guía que conozca la región, y con un equipo de trabajo que
cooperate with the analysis and comments, providing explanations when necessary;

☐   mantener orden y buena conducta en todos los recorridos;

☐   observar las normas ambienta (proper waste management, noise, etc.) and
due respect for places and people;

☐   motivar permanentemente a los participantes para que observen y anoten
the most important details

☐   escuchar atentamente las preguntas e incentivar los discussions and
exchange
of participant information

☐   analizar en detalle los casos críticos que se presente, promoviendo la search for on-site solutions

☐   estimular a los participantes para que pregunten a las personas del lugar los details you need to know

ELEMENTS THAT HELP US TO KNOW THE COMMUNITY

"It is always opportune to remember that in the study of communities we cannot limit ourselves only to structural, physiological and demographic factors, but it is necessary to investigate the social problems of the community, often resulting from conflicts between attitudes, values, personalities, institutions and economic, racial, religious, political and cultural groups".

It should not be left aside:
*   the child's adjustment in society;

*   assimilation of the national immigrant or foreigner;

*   penetration of social, moral and religious values;

*   the oscillations of economic and political events and their social repercussions;

*   group formation, segregation and racial discrimination.

*   the emergence of smaller communities within the local community;

*   marginalized groups;

*   situations of critical poverty;

*   the place of women;

*   risk groups;

*   existing conflicts;

*   prejudices; etc.

WHAT DO WE SURVEY?

In projects aimed at the education of adolescents, young people and adults, it is necessary to contemplate the survey of:

- Needs and problems

- Demands, interests and expectations

Available resources

Concepto de comunidad

Bernard (1973)
Comunidad simbólica que incluye lazos emocionales comunes, compromiso, cohesión social y continuidad temporal.

Marchioni (1990)
Conjunto de personas que habitan el mismo territorio, con lazos e intereses comunes, compuesta por: territorio, población, demanda y recursos

Hillery (1995)
Localidad compartida donde existe interacción social, relaciones y lazos comunes.

Ezequiel Ander Egg (1980)
Unidad social, cuyos miembros participan de algún rasgo, interés, elemento común, con conciencia de pertenencia, situados en una determinada área geográfica en la cual la pluralidad de personas interactúan más intensamente entre sí que en otro contexto

CEC (2004)
Comunidad como grupo social. Cualidad de convivencia. Relaciones que se construyen en medio de procesos de participación, cooperación y proyecto.

When the educator comes into contact with the needs of the inhabitants of the area where he/she will develop his/her educational work, the first data collected will allude to their needs and problems. When observing or receiving comments from the people, with reference to the reality in which they live, he/she will try to capture the dimension of these situations by trying to transcribe verbatim the main ideas, taking into account the techniques and instruments of diagnostic evaluation that he/she decides to use (interviews, surveys, field diary, anecdotal records, etc.). The decoding task will be carried out in a timely manner with the students, once the class group has been formed.

If the educator does not belong to that place and does not know the sociocultural characteristics (among others, the restricted codes of language), and its social dynamics, it is very possible that those ideas and spontaneous and personal or collective expressions, those people, contaminate with prejudices, interpretations or personal evaluations of the educator, thus losing the possibility of carrying out an educational process that favors the awareness for the transforming action on that reality or situation that the learner lives as conflictive.

The degree of problem solving will depend on the characteristics of the problem3 and the sectors responsible for or involved in the task.

"It is important to emphasize that the adult educator has a fundamentally formative mission with social intervention. Therefore, their function in the face of problems related to Unsatisfied Basic Needs presented by the learners is eminently educational.

Usually, the educator, moved and concerned about the solution of vital needs of the student and the community, assumes the role of Social Worker and the responsibility of responding to issues that the educational service cannot solve because they are not within its competence.

"It is very likely that the educator perceives needs that the learner does not feel as such. Hence the importance of the purpose of this stage: the collection of data that will be recorded as faithfully as possible.

Let us remember that both in the literacy process (in the recording of the Vocabulary Universe and in the application of the literacy methodology) and in the implementation of Vital Learning Units, the educator and the group develop the teaching-learning process around needs, demands, problems and interests. Therefore, it is important to take into account the following conceptualizations:

Need: is a lack or state caused by a deprivation. It is also called mot i- vo or internal variable since it is a cause that moves towards an action with the purpose of satisfying or eliminating it.

This concept has different scopes:

1. Basic or Primary Need: indispensable for survival (e.g., food). Also called vital needs. This category includes Unsatisfied Basic Needs (UBN).
2. Accessory Necessity: Not indispensable for survival (e.g., collecting).
3. Felt Need: is that which the group or individual perceives, identifies or recognizes as a lack.
4. Manifest Need: it is patent; ostensible. It is that which is exposed; in sight.
5. Potential Need: is the one considered as possible.
6. Latent Need: it is hidden but exists; it is not manifested or externalized.

It is important to keep in mind that the demand is the expression, that the group or the individual makes, of the need. This concept is associated with the claim, request or petition.

The problem represents an issue, circumstance or set of facts that hinder the achievement of some purpose or goal that can be solved. It admits verification with a certain degree of probability; verification in the present and application of known procedures for its resolution.

2- HOW DO WE DO IT?

Through:

*   An approaching tour.

*   Bibliographic consultations.

*   Map revision.

*   Development of a systematic observation plan.

*   Development of cooperative work.

*   Search for community informants.

*   Organization of pro-enrollment events.

*   Recovery of instances of practice and reflection on practice to arrive at "knowing how to do with knowledge and awareness".

Let us briefly analyze each of the above situations:

AN APPROACHING TOUR
This first contact with the community should be as comprehensive and inclusive as possible.
Field work, visits to students, and the placement of posters in businesses to advertise the educational offerings provided by the center or adult school provide us with interesting information.
But we must work on extending these tours to other areas of the place in order to compare and recognize the center and the periphery; factory centers, institutions, etc. It is also possible to talk to the villagers in order to collect more precise data. Adult education takes place within the framework of institutions, and in this approach, the objectives, histories, users, problems, concerns, aspirations, etc. should be internalized.

BIBLIOGRAPHIC CONSULTATIONS.

This section is particularly interesting to rethink ourselves as educators in a given community. Many times, we ignore the amount of bibliographic sources that speak and were written about the places where we work.
Even the notebooks of learners who have already graduated, the previous diagnoses, the archives of the institution's events, give us an idea of how Adult Education is understood and what value is assigned to it.

We can consult:

- Historical sources.

- Statistical sources (municipal, provincial, school, etc.).

- Memories of institutions.

- Institutional or settlers' archives.

- Local or zonal newspapers, mass circulation newspapers, etc.

- Family and institutional photographs and albums.

- Recordings.

- Other sources.

At this point it is important:

Do not forget the purpose of these consultations.
- Look for reliable sources. Analyze them critically, asking yourself, for example, if they are not intentional data, in the case of statistics.

- Questioning the competence of the authors, etc.
- Working with them before and during the diagnostic process. As we all know, the diagnostic stage implies an educational process per se.

MAP REVISION
Maps give us an idea of how man relates to nature and the challenges it poses. Inscribes the educational service in larger frames: the locality, the region, the country, etc. It can give an idea about how others see the area. The most important are:

- the political map;

- the physical map;

- ethnographic map and,

- the communications network map.

DEVELOPMENT OF A SYSTEMATIC PLAN FOR MONITORING

The field to be discovered is complex and the subjects who enter it to observe it are traversed by a series of prejudices and ideologies. But observation must be as comprehensive and as close to reality as possible. For this reason, the elaboration of a plan is a help and an imperative. Deciding with an experienced member what to observe, determining the instruments, the times and the terrain, ensures better control of this moment.

Keeping the elements observed in writing and not relying on the observers' memory helps to make the most of the information.

DEVELOPMENT OF COOPERATIVE WORK

We all know, from personal experience, that group work is productive, but that it requires practice and a series of attitudes.

We also know that not everything can be done in groups. In the case of diagnosis, forming a work team is the optimal mechanism to build the scaffolding for real participation.

Therefore, group work must be learned and Adult Education is the ideal and appropriate place for it.

In a group, each individual brings his or her personal knowledge, skills, motivations, and failures or imaginary ideas about the collective work.

The adult attends class after a day's work and expects to find productivity in the group; he does not want to waste time.

Training in the group task is costly; therefore, the educator must take certain precautions so as not to discourage his learners in this difficult task:

- Plan in a realistic, progressive and flexible manner.
- Aims and objectives should be clear and determined by the group.
- The classroom climate should be experienced as harmonious; the members will agree and reach a consensus on a series of questions beforehand, trying to achieve a common vision regarding the task to be undertaken.

- Highlight achievements and results in a tangible and constructive way.

SEARCH FOR COMMUNITY REFERENTS

Community referents should provide information related to culture and other educational agencies in the area.

In addition, we must start from the fact that in all communities there are previous histories that refer to Adult Education, whether formal or informal. The presumption of being the founders of Adult Education in the place must be cleared and we must start to track down "the memory people" so that they can provide us with useful data.

You can visit or consult the people listed below who will certainly provide relevant information:

- educators and/or literacy educators;
- experienced on-site staff and technicians, living on or off-site;
- neighborhood leaders and leaders;
- church representatives; among others.

Data will be collected by means of:

- interviews;

- questionnaires;

- surveys.

The information collected is decoded and given relevance by means of:

- data tabulation;

- the relationship and analysis of the content of the information;

- problematization.

BIBLIOGRAPHY

BRANDAO, Carlos. Pesquisa participante. Brazil, Ed. Brasilense.
DEMO, Pedro. Participant research. Myth and reality. Bs. As., Kapelusz. FERREIRA, Francisco de P. Teoría social de la Comunid ad. Spain. Ed. Católica.
FERREYRA, Erasmo N. La Lenguajización para la Educación de Adultos . Bs. As., Genitrix. FILCAR. Buenos Aires and Conurbano Bonaerense Street Guide. Bs. As., Ediciones Filcar. FREIRE, P. Raising awareness in the rural world: Extension or Communication? Bs. As. Siglo XXI. GAJARDO, Marcela. Evolution. Situación actual y Perspectivas de las estrategias de investigación participativa en América Latina. Santiago de Chile, FLACSO.

INDEC. Population Census 1991. Bs. As., INDEC.

IOVANOVICH,M. L. Technical Circular. Diagnosis. La Plata, DEA. and FP.
IOVANOVICH, M. L. yALURRALDE, E. M. El Diagnóstico Integrador Participativo - Proyectivo: Un salto a la Transformación de la Educación entre Adultos. Bs. As., Dunken.
IOVANOVICH, M. L. Jurisdictional Project of the Federal Program of Literacy, Basic Education and Work for Adults of the Municipality of the City of Buenos Aires .

LE BOTERF, Guy. Participatory Survey. Brasilia, UNDP - UNESCO Project.

MOSER, Heinz. Action-research as a new paradigm in the Social Sciences.
NAZAR, Ángel. Acción cultural como estrategia de desarrollo. Bs. As., Plus Ultra.
OQUIST, Paul. La Epistemología de la investigación - acción. Quito, United Nations.
SALINAS, Willy B. The Survey - Participation. Lima, Asociación de Asistentes Sociales de Perú. SIREAU, Albert. Esbozo de una matriz de interrelaciones. Un instrumento de análisis interdisciplinario y de programación intersectorial. UNESCO - OREALC.
SPENCE DONALD,P. NarrativeTruth and Historical Truth: Meaning and interpretation in Psycho- analysis. New York, Basic Book.

GLOSSARY:

Basic Concepts

Sustainable agriculture, a system of agricultural production that allows stable production in an economically viable and socially acceptable way, in harmony with the environment.

Protected areas, specific parts of the national territory declared in accordance with current legislation, of ecological, social and historical-cultural relevance for the nation, and in some cases of international relevance, especially devoted, through effective management, to the protection and maintenance of biological diversity and associated natural, historical and cultural resources, in order to achieve specific conservation objectives.

The competent authority is empowered to apply and enforce compliance with the provisions of this Law and its complementary legislation.

Environmental cost is the cost associated with the current or prospective deterioration of natural resources.

Environmental damage, any significant loss, diminution, deterioration or impairment of the environment or one or more of its components, which occurs in contravention of a rule or legal provision.

Sustainable development, a process of sustained and equitable improvement in the quality of life of people, through which economic growth and social improvement are pursued in a harmonious combination with environmental protection in such a way that the needs of present generations are met without jeopardizing the satisfaction of the needs of future generations.

Hazardous wastes, those coming from any activity and in any physical state that, due to the magnitude or modality of their corrosive, toxic, poisonous, explosive, flammable, biologically pernicious, infectious, irritating or any other characteristics, represent a danger to human health and the environment.

Radioactive wastes, those that contain or are not contaminated with radionuclides found in concentrations or with activities above the levels established by the competent authority.

Biological diversity, variability of living organisms from any source, including, among others, terrestrial and marine ecosystems and other aquatic ecosystems and ecological complexes of which they are part. It includes diversity within species, between species and of ecosystems.

Ecosystem, complex system with a certain territorial extension, within which living beings interact with each other and with the physical or chemical environment.

Environmental Education, a continuous and permanent process that constitutes a dimension of the integral education of all citizens, oriented to the acquisition of knowledge, development of habits, skills, abilities, capacities and attitudes, and the formation of values, to harmonize the relationships among human beings and between them and the rest of society and nature, in order to promote the orientation of economic, social and cultural processes towards the SD.

National Environmental Strategy, expression of the Cuban environmental policy, in which its main projections and guidelines are expressed.

Environmental impact study, a detailed description of the characteristics of a work project or activity to be carried out, including its technology, which is submitted for approval within the framework of the environmental impact assessment process. It must provide a well-founded background for the prediction, identification and interpretation of the environmental impact of the project and describe the actions that will be taken to prevent or minimize adverse effects, as well as the monitoring program that will be adopted.

Environmental impact assessment, procedure that aims to avoid or mitigate the generation of undesirable environmental effects, which would be the consequence of plans, programs and projects of works or activities, through the prior estimation of the modifications to the environment that such works or activities would bring about and, as appropriate, the denial of the necessary license to carry them out or its concession under certain conditions. It includes detailed information on the monitoring and control system to ensure compliance and the mitigation measures to be considered.

Environmental management, a set of activities, mechanisms, actions and instruments aimed at guaranteeing the administration and rational use of natural resources through conservation, improvement, rehabilitation and monitoring of the environment and control of human activity in this sphere. Environmental management applies the environmental policy established through a multidisciplinary approach, taking into account the cultural heritage, accumulated national experience, and citizen participation.

State environmental inspection, activity of control, inspection and supervision of compliance with the legal provisions and regulations in force regarding environmental protection, with a view to evaluating and determining the adoption of the pertinent measures to ensure such compliance.

Environmental license, official document, which without prejudice to other licenses, permits and authorizations that in accordance with current legislation must be granted to other state bodies and agencies, is granted by the Ministry of Science, Technology and Environment to exercise due control of compliance with the provisions of current environmental legislation and contains the authorization that allows a work or activity to be carried out.

Environment, system of abiotic, biotic and socioeconomic elements with which man interacts, while adapting, transforming and using it to satisfy his needs.

National Environment and Development Program, a concrete projection of Cuba's environmental policy, which contains guidelines for the action of those involved in environmental protection and for the achievement of the SD. It constitutes the national adaptation of Agenda 21.

Marine resources, the coastal zone and its protection zone, bays, estuaries and beaches, the insular platform, the seabed, and the living and non-living natural resources contained in maritime waters, seabed and subsoil, and emerged zones.

Natural resources, all components of the environment, renewable or non-renewable, that satisfy economic, social, spiritual, cultural and national defense needs, guaranteeing the balance of ecosystems and the continuity of life on earth.

Landscape resources, geographic environments, whether surface, subway or underwater, of natural or anthropogenic origin, that offer aesthetic interest or constitute characteristic environments.

National System of Protected Areas, a set of protected areas that, in an orderly manner, interact as a territorial system that, through the protection and management of their individual units, contribute to the achievement of certain environmental protection objectives.

Environmental variable, an element of the environment that can be measured or evaluated by different qualitative or quantitative methods.

Index